D1645894

PACKED

THE
FOOD ENTREPRENEUR'S
GUIDE:

HOW TO GET NOTICED AND HOW
TO BE LOVED

Tessa Stuart

Published by Stamford Brook Press

ISBN: 978-0-9576028-0-9

TABLE OF CONTENTS

INTRO

WHAT THIS BOOK WILL DO FOR YOU

Do you have an idea for a unique and fabulous food or drink business?

While supermarket shelves are packed with mass-produced and low-quality products, there's an ever-widening gap for brands that have a bit more to give and a bit more to say for themselves.

If you want to succeed – if you want your product to fly off the shelf – you need to pay attention to branding. The product itself can only get you so far.

And here's where this book comes in. It'll help you figure out your next steps to turn your product into a brand that can compete with the Branded Big Guns in the supermarkets.

You don't need millions; you just need proven strategies, creativity and determination.

The inattention problem

In the aisles, we're on autopilot.

We zombie along, sticking to familiar brands and products (Heinz, McVitie's, Cadbury...) unless completely won over by enticing packaging.

As a newer product with an unfamiliar brand name, you'll struggle to get noticed.

So, how do you compete?

What this book will do for you

Your product needs to stand out and be remarkable if it's going to compete with the big brands and get bought by customers.

By using the recommendations and advice in this book – based on my 15 years in the food branding industry – you'll be able to make that happen.

You'll also find lots of advice from the brilliant food and drink brands I know and have worked with: Gü Puds, innocent drinks, Muddy Boots Foods, Rude Health, MOMA!, Higgidy, Graze, G'NOSH, Jimmy's Iced Coffee and more.

This is a tips book – to pick up, browse, and give you some inspiration and ideas. You'll find cracking ideas on:

- Identifying your niche

- Testing your market

- The right pack sizing

- Clever packaging

- WOW! design

- Defining YOUR difference

- Social media sparkle

- Super sampling in-store

- Dynamic distribution

Everything in this book has been proven to work.

That's not to say it'll be easy – it takes grit, graft, guts and a great product to be a successful food or drink brand.

But this book will keep you on the right track, give you the right goals to be aiming towards, and tell you what works and what doesn't when it comes to packaging and positioning your product.

WHO I AM AND WHY I WROTE THIS BOOK

I used to run focus groups

And what surprised me when I started was how much people tried to project a much better version of themselves, or say what they thought wanted to be heard.

Unfortunately, that isn't very helpful for the food and drinks companies that rely on our input – especially small, unique companies that need real customer insight to help them compete with the big boys.

Now my time is spent in the aisles...

Where I have real, live, funny and illuminating conversation with people as they scoot through their shopping in the way that we all do – distracted, in a hurry, not really thinking that hard.

Quick, candid conversations. Not contrived ones.

I peek into people's trolleys to see how they balance health and convenience when they shop. And I ask them to tell me about their real shopping choices from the shelves – not recall it for me in a faux-memory reconstruction in a hot stuffy sitting room with eight total strangers.

I get real insights from real shoppers

The vast majority of shoppers are really interested in being part of a quick process about what they choose to buy. They happily show me what works for them and what doesn't. They'll explain how a product feels to touch, or look at, or hold. And they'll tell me what they know about the brand, the ingredients, the price and the competition – everything that helps them decide whether to buy the product or not.

Talk to 40 or 50 people this way, and you can map a whole audience for a product. And on top of that, you can map all the "trigger points" that bring people to a brand.

Once you've got all these insights, you'll know what needs to be changed or fixed (the pack shape and design, pricing, messaging, ingredients, story…), to get more people buying.

And I do mock-ups of new products too…

I started doing mock-ups when innocent drinks wanted to test the pack for a new product in their range. It worked a treat: the mock-ups flew off the shelves (and I had to run after shoppers to explain why they couldn't be bought!).

Since then, I've done mock-ups plenty of times – they're invaluable for indicating whether the product is tempting to shoppers before all that money gets wasted on pack design, production, and so on.

I'm evangelical about all this...

This "in the aisles" approach is the best one. Researching in the aisles helps you understand how a product *really* talks to shoppers – people who're harassed, hurried and not trying to impress anyone else.

I arm the little brands with all the ammunition they need

For the "little guys" to compete with the big food companies (who have huge market research budgets and an established brand name already), you need honest and reliable feedback from shoppers. That's what I do, and as my many clients will attest, I do it well. (To see their testimonials, and for more information on what I do and how, visit **www.packedbranding.co.uk**.)

It's a big ol' battlefield out there, but the Davids CAN win against the lumbering Goliaths.

SECTION 1: GET YOUR PRIORITIES STRAIGHT

TASTE IS MORE IMPORTANT THAN ANYTHING

Branding is extremely important – but there's no point having a riveting brand story, fantastic packaging, beautifully crafted words and an amazing point of difference if your product's no good.

Richard Reed of innocent drinks explains:

> "Product is king. There's no point having an idea but a lousy execution. We had a strong idea after spotting a gap in the market. But the product had to match. If it didn't taste good, it wouldn't have worked. We didn't invent the smoothie – we just made a better one that tasted better and was more nutritionally loaded. That's what you have to do – be better than your competitor."

If people aren't wowed – if they aren't saying "That's amazing!", "Mmmm... can I have some more?", or "Utterly delicious" – GO AND START AGAIN.

Make it taste better.

Do product testing in offices (focus groupees are paid to tell you what you *want* to hear), and listen out for red flags:

"Hmmm... it's quite nice."

"I guess I might buy it occasionally."

"It tastes just like…"

Those are the kiss of death. Not good enough! Your product won't get you remembered, bought or recommended.

Taste is your MAJOR WEAPON against the competition.

A great-tasting product means you can start making people aware of you and getting them to remember you BEFORE you have a big marketing budget or lots of staff.

Your product needs to look great and have strong brand messaging to stand out on the shelves. But it needs to taste damn fine to get talked about and bought again.

When you taste terrific, people will go out of their way to find your product again.

They'll enthuse about you, recommend you, and introduce you to their friends.

It's how innocent drinks became popular in the early days: the drinks were delicious enough that people told their friends.

"Good enough" is NOT good enough.

Sales will plateau – however great the branding – when buyers realise the product just doesn't taste incredible (or tastes just like something else that's cheaper).

If you don't WOW, your sales won't grow.

FOR LOVE OR MONEY?

What motivates you?

Are you driven by the need to get up in the morning and bake?

Or make fruit-flavoured ganache?

Or blitz soup?

Or press fruit?

Or create dips?

Or are you driven by the thought of selling out in ten years' time to a bigger food company?

Attracted by the glamour of being a food and drink entrepreneur?

Are you planning that exit route as you begin?

Whether you're in it for the love or the money, you need to watch the cash flow.

Artisans are sometimes oblivious to profit.

Entrepreneurs are sometimes too much in love with it.

Artisans can lose the opportunity to build their brand and make a living by doing what they love. Entrepreneurs can miss the chance to attract the right sort of (affluent) customers because they don't pay enough attention to quality.

You should aim for a balance – where quality isn't sacrificed for profit, and your profit margin doesn't dictate the quality.

WHAT ITCH ARE YOU SCRATCHING?

Take your idea and answer this question from Richard Reed of innocent drinks:

"What itch are you scratching for your customer?"

Is anyone else currently scratching that itch? That's fine.

But are you GENUINELY BETTER or MORE CONVENIENT than the existing itch-scratchers?

In what way? And for whom? You need to be genuinely better or more convenient *for someone.*

If you're not already, you need to find answers to these questions: *how* are you better, and *who* are you better for?

Who do you want your customer to be?

Phil Pinnell, the founder of Scratch Meals, was sick of ready meals and wanted to cook from scratch in the same time it takes to heat a ready meal. He solved his OWN problem – but before he turned his solution into a business, he made sure it was a problem others were having too.

Do you think others share your problem?

If you come up with *another* smoothie or *another* cupcake, make sure it solves a *genuine problem that people have with existing smoothies or cupcakes*. Not everyone has to have the problem, but enough do.

You need to create their Holy Grail.

You need them to tell their friends about you.

And when they tell their friends about you, what do you want them to say? Think about how you want to be talked about.

TIPS FROM AN EXPERT: THE ELEMENTS OF SUCCESS

Ben Jones is the co-founder of Graze – the fastest-growing start-up food company in the UK, brilliantly sold online to desk-bound office workers via a very clever subscription model.

Here are his elements of success:

Spiritual

Good things happen to good people. In any start-up you always need a little bit of luck – so be nice to everyone and boost your karma.

Theoretical

Know your costs. Cash flow is key, and you don't want to start anything without understanding its sustainability. Once you're up and running, you've moved to the "actual", so make sure you constantly review and update the black and white, as this is the business in its purest form.

Magical

Always keep innovating and do it in every part of the business – from your operations to the products and services. There's a lot of competition out there, but if you're innovative, your customers will benefit directly, they will love it, and you will make yourself proud.

Physical

Surround yourself with the best people you can find. These people will share the ambition and passion you have for the company, be smarter than you at what they do, have a positive impact on the company's culture and motivate you to be better.

Practical

Don't be afraid to take risks. Use your common sense and be yourself.

LISTEN TO PEOPLE IN THE KNOW (AND IGNORE EVERYONE ELSE)

Belinda Williams of Yorkshire Provender stresses:

> "Don't be persuaded by professionals that they know better than you. Follow your instinct and stay confident. Just because someone has been involved in success does not mean it was their own."

Jim Cregan of Jimmy's Iced Coffee says:

> "Squeeze every drop of information from people in the know who have time for you: trends, finance, distribution, marketing, logistics, anything. Even people you don't know, ask them! Knowledge really is power!"

But make up your own mind for your own business.

It's your business, not anyone else's, growing and developing uniquely along with you.

IF YOU'RE HIGH-QUALITY, THEY CAN'T TOUCH YOU

If your product is high-quality, it'll be much easier to retain your market position when cheaper alternatives crop up.

As Camilla Barnard of Rude Health says:

> "Make a good product that really is distinguished by quality and difference. Innocent has never really been threatened by supermarket own-label products because of their adherence to those two principles."

Richard Reed of innocent drinks says:

> "It's not that we never had competition. All the big companies have tried, but none of their products ever worked because they compromised on the ingredients to be more profitable. The profit margin we make is so much less than a big company would make. As soon as you start putting sugar or preservatives into a smoothie, then you've broken the promise of what a smoothie is. It makes sense on the spreadsheet, but not in real life. A smoothie has to be natural, delicious and healthy, or people aren't going to buy it."

TIPS FROM AN EXPERT: STARTING OUT

James Averdieck is the founder of Gü Puds.

Here are his tips:

Equity is precious

If you're considering funding/incentivisation, don't give it away easily just because your business is worth nothing at the start. One day, if your business is successful, 1% of it could be worth £300k – which puts it into perspective.

Have a small company mentality – even when you get bigger

Small companies should focus on their strengths, which is being small, dynamic and flexible. You don't need lots of fancy expensive marketing or lots of employees. It might be tempting – especially if lots of "experienced" people are advising you to scale the business. But do as much yourself as possible for as long as don't get too stressed!

Don't stop innovating

Even when you've got a product, spend lots of time on innovation. Look out for clues in other product areas the whole time and make links. Trust your gut instinct over research, but remember to sleep on big decisions!

Leave aside time for fun

Remember to relax. Think about the reasons you wanted to have your own business in the first place. I bet it was all about freedom and independence, and not about money. So make sure you enjoy the freedom.

Don't think about the "exit"

Don't waste time thinking about your exit strategy. If someone wants to buy your business, they'll approach you. Then, if you're interested, the best thing to do is hire an agent and get an auction going.

BE TRANSPARENT ABOUT YOUR PRODUCT AND ITS INGREDIENTS

Richard Reed of innocent drinks explains why:

> "Our business only exists because people buy the products. It's not like they're made to. The consumer has the absolute right to know everything – they should know exactly what's in the product, and where it comes from."

Your ingredients tell your story. If you use Shipton Mill flour to make your crispbreads – flour that's been stone-ground – then say so. If you use only Michael Cluizel couverture single plantation chocolate, then shout it out – because to a chocolate-lover who knows Valrhona or Amedei, it's important.

TIPS FROM AN EXPERT: THE FUNDAMENTALS

James Foottit is the co-founder of Higgidy Pies.

He says:

1. Look for advice from other suppliers. We've been amazed how other suppliers are ready to offer advice. Don't be afraid to ask: everyone loves being asked for advice and you need some friends to negotiate the challenges.

2. Perfection is the enemy of the good. Don't wait until you have perfected every aspect of your idea – get started and get better as you go along.

3. Know your costs. When we started out we often underestimated the real costs of things like distribution.

4. Be disciplined about financial reporting. You need to know how good (or in our case bad) the numbers are. You need to be reviewing your numbers on a monthly basis.

5. Have fun; work with people you like. You're probably going to put some long hours in, so make sure you're enjoying it.

NOBODY CARES AS MUCH AS YOU

Camilla Barnard of Rude Health explains why:

> "We were confident that we knew cereal right from the start but felt like rank amateurs when it came to organising events, designing exhibition stands and most of the things that you need to do to get cereals to the people who eat them.
>
> So it was very tempting to let specialists offer their services.
>
> Now we do almost everything ourselves. I think you can interpret this in two ways:
>
> 1: We are total control freaks who can't delegate anything.
>
> 2: Real enthusiasm and commitment are as important as expertise, and no one is more committed to or enthusiastic about your business than you."

TIPS FROM AN EXPERT: WHAT INVESTORS WANT FROM AN ENTREPRENEUR

Charlotte Knight of G'NOSH shares her list:

What's your "unfair advantage"?

Make sure your idea is solving a problem, addressing a real opportunity and not just developing an idea. It must have a strong, unique business proposition. Know who you're up against – direct competitors as well as substitutes – because it's important you have an edge over the competition.

Be tax efficient

Get EIS or SEIS approval from HMRC. No red tape is involved and it makes your business significantly more attractive to investors as a tax-efficient scheme.

Have a "story"

And pitch it punchily. Support your story with a business plan as a back-up. Most investors will want to learn about YOU initially. Remember: people buy from people, and they want the business to be supported by an excellent management team. The investment pitch is crucial in establishing your charisma and likeability. Keep it professional, knowledgeable and convincing.

Know your "commercials"

Invest in business planning and thoughtful financial projections. Know your end price point to the customer, your distribution model, your total cost of goods, and your margin aspirations.

What's your exit route?

For investors to really be interested, there must be a clear exit route for the business.

Be driven

You need to have a stack of passion, drive and networking ability to lead and execute against your plan.

(If you want more help with manufacturing or investment advice, contact Charlotte Knight at G'NOSH.)

SECTION 2: DESTINATION SUPERMARKETS

AIM FOR THE SUPERMARKETS

The margin and costs involved in supplying a supermarket are very different from supplying a few independents.

There are many benefits in sticking to independent delis. For example, there's less pressure to scale up, and you'll have a more personal relationship with the deli owners. You might not want to ever consider scaling up if your product is exquisitely handmade, and you are a kitchen table business.

But this book is about aiming high – towards the supermarkets – so you'll be able to reach more people in a day than you ever could via the independents.

THE ROUTE TO THE SUPERMARKET AISLE

The usual path is:

- Market stall

- Local delis

- Whole Foods/Planet Organic/As Nature Intended

- Selfridges/Harvey Nichols/Booths

- Ocado

- Waitrose in a restricted range of stores (maybe a local listing)

- Co-op

- Waitrose in a bigger number of stores

Followed by listings in:

- Sainsbury's

- Tesco

Some niche businesses do leap from independents straight to Ocado and do very well, as COYO coconut yoghurt has done.

WORK BACKWARDS FROM YOUR BUSINESS MODEL

The UK food market is the most competitive and innovative in the world.

If you want to get on the shelves of the multiples, you'll need to

either:

create an entire category (as innocent drinks did with their smoothies)

or:

rejuvenate a category (as GNOSH did with dips).

So when you plan, track it **backwards** from that point.

TIPS FROM AN EXPERT: FINDING A MANUFACTURING PARTNER

Charlotte Knight is the founder and owner of G'NOSH dips. Here she shares her list on what you should be looking for when choosing a manufacturing partner:

Strategic fit

Long-term benefit should accrue for both your business and the supplier.

Capability

Check their accreditations to ensure they'll manufacture the products safely and to high quality. Ideally they should have a track record in retail supply – it'll help when you have to supply directly to UK multiple grocers.

Capacity

They need to be able to respond to seasonal or promotional peaks in demand and capacity as the brand develops.

Competitive pricing

You can check that they're giving you good prices by doing your own research on raw material costs – that way you'll have a benchmark to work from.

Desire

They should *want* to work with you – it's key to building a trusting, long-lasting relationship.

GET YOUR PRICE AND MARGIN RIGHT FROM THE START

Raw material costs, production costs, and distribution take chunks out of your profit.

That's before any premises costs, employee costs, marketing spend, salary for you, or investment in research and development for your next product or flavour variant.

And it's before the supermarkets tell you they want X margin, and want to give your product away on promotion. (And they WILL want to give your product away, so be ready.)

So do the maths at the start and get your bottom line where you want it.

I can't give you a template for getting it right, because it'll depend on how "premium" you want to be, your raw material costs, your production and distribution issues, the category you're entering, and how price sensitive the shoppers in your category will be.

But work on your own figures and get everything right.

Because if you get it wrong, it's very hard to recover from: if you've already launched and you then need to fix your bottom line by adjusting your pricing or sizing, you'll get squeals of outrage from customers who've got used to spending a certain amount.

As Giles Brook of Bear Nibbles says:

> "If your bottom line isn't where it needs to be before launch, don't launch until you know you can categorically get there."

TIPS FROM AN EXPERT: PRICING YOUR PRODUCT

Camilla Barnard is the co-founder of Rude Health.

Here is her key tip on price:

Discover your price point

"Find out what the price point is that a supermarket wants for your product, and the price point that works for you and the consumer. You can go to see a supermarket buyer for advice – you'll need to supply them with a product, a clear idea of price, and branding."

Here are my other tips on pricing:

Charge a premium for high-quality products

If your product has lovely ingredients that have been thoughtfully combined in a slow food process, price it well.

People in Chiswick (West London) don't baulk at an artisan bread loaf from The Slow Bread Company at £4.50 a loaf. Sold through speciality delis, people like treating themselves to it on weekends. The same bread can be found in my lovely Hammersmith gastropub, which prides itself on sourcing locally. I'm happy to have it along with a half of Adnams Broadside or a Fullers London Pride. See how context is crucial to self-treating spending decisions!

It's a lot easier to come down in price than go up

So if anything, price higher. But don't price too high, because that'll rule you out for too many people. (Yep, this decision is complicated!)

Pricing higher is particularly important at the start, because you'll need every margin point you can squeeze out of your product – especially if you're a start-up or small product company.

At the start, without the benefits of big volumes of products produced and sold, you need to price to cover your costs. As you grow, you'll get economies of scale on production and packaging – which means you can negotiate costs down on these.

And if possible, try to price so that you can cover yourself from having to make any price increases for the next few years. That way, you can absorb ingredients costs for some time – instead of passing them on to customers. (Alternatively, you could buy cocoa futures.)

Innocent drinks got hit in 2008 by rising fruit costs. And at the same time, foreign exchange movements were working against them AND there was a consumer slow-down. It was a tricky time for innocent, who had to make staff redundant. Protect yourself in case such an unfortunate series of events should ever happen to you!

THE KEY QUESTION TO ASK ABOUT RATE OF SALE

Miranda Ballard of Muddy Boots Foods makes burgers and meat loaf, stocked in Waitrose and on Ocado.

She explains it:

> "We were pretty good at having the audacity to call up successful food companies a few years ahead of us and saying, 'Hello, we want to be a food company too… could we buy you a coffee and pick your brains?'. We were amazed and delighted at how many were happy to help, and we really learnt a lot. (We still do this now; I don't know why we'd ever stop.)
>
> However, the one question we never asked – which would have been the MOST important – is very simple: 'How many units do you sell per store per week?'
>
> We imagined that the big food companies sold hundreds of units per supermarket store per week. And from that, we imagined we'd sell a lot less. But we still wildly over-estimated – surely 30–40 units per week, that's only 10% of what the big boys are selling, surely…
>
> Actually, in our category, only the big companies get near to 30–40 units and even then, it's probably when they're on special offer. We're more like 9–12 units per

week on average and if we'd have known that before we started with supermarkets, our first cash-flow would have been more realistic."

OCADO OPPORTUNITIES

Ocado is a brilliant shop window for smaller producers.

It gets you right to the desktops/iPads/iPhones of the time-poor, cash-rich working woman.

Ocado like different, original, "challenger" brands, because they want to differentiate themselves from the other supermarket online services. They want to show they're a *better* alternative to Waitrose.com, Sainsburys.com, and all the others.

So grow your business with them; they need you to be available through them exclusively.

And pay special heed to online customer reviews – which can give you a surprising early and accurate warning of any failings in your production kitchen. (Ocado does give you a right of reply.)

Cross-promote with other brands on Ocado to increase Ocado's basket size – Ocado likes that. Hook up with other brands that are complimentary to you and suggest that you do a joint promotion together, as G'NOSH and Popchips do. Dips and crisps: a natural combination.

TIPS FROM AN EXPERT: GET YOUR NAME OUT THERE

Alastair Instone is the founder of School of Food – a travelling cookery school in London. He knows exactly what it takes to get your brand name out there.

Here are his tips:

Your product needs to be fantastic

For every hour you spend on getting your name out there and promoting your product, spend two hours on ensuring your product is fantastic. If it's a let-down, customers will only buy it once and all that marketing effort will be for little reward. If the product is fabulous, they'll buy it repeatedly and tell all their friends. Soon you won't be worrying about how to get your name out there – you'll be concerned with coping with all your new customers.

You don't need to use every marketing technique!

Find one or two marketing techniques that fit your budget and product, and concentrate on them. You don't need to use all the techniques out there, and you don't need to invest in the latest flashy way to attract customers! For School of Food, two approaches were enough to get off the ground and remain valuable: make the business's services easy to find on search engines; and encourage happy customers to spread the word about the service by giving them discount codes to pass on to their friends.

If it's not working, change tack

Be brave and change tack if a particular marketing technique isn't working (and don't carry on spending money on something that's not working in the hope it'll magically improve). In the early days of School of Food, some positive press coverage was enormously valuable – it raised the business's profile and credibility, and brought in lots of new customers – but the return diminished surprisingly quickly, and other activities soon became more cost-effective. But, because it worked so well at the start, I carried on putting time and money into PR.

You only need as many customers as you can cope with

If you invest in marketing but can't produce enough for all your new customers, that's not successful marketing – it's a waste of time and money.

IF YOU DON'T WANT TO AIM FOR THE SUPERMARKETS...

You might prefer to stick with the independent stores, or sell your product directly to customers through your website. That's fine – most of the information in this book still applies to you.

Here's some extra information just for you:

About selling directly to customers through your website

Selling directly to customers works particularly well if...

- You're a small-scale speciality product (and won't ever be able to scale too much), like a rare tea or a speciality chocolate maker

- You have a relatively long shelf life

- You're easy to post (e.g. tea)

- Your remote, artisan, hedgerow-picked and foraged location – which prevents you from having access to supermarkets – is part of your appeal (e.g. JAMSMITH)

- Your postage costs are proportionate to your price.

If you *do* decide to sell directly to customers online, consider a subscription model – it'll cut your costs of admin and provide you with cash flow and predictability of work.

About being stocked in independent stores

Dan Shrimpton, co-founder of Peppersmith, deliberately set out to find other distribution channels for his mints and chewing gum – aiming for cafes in particular. Here he shares his advice:

- You need to know from the start where you're aiming for: the impact on margins and distribution is fundamentally different in an independent compared to a supermarket.

- Where you're stocked is a key part of your brand. If a consumer discovers you first in a local deli, it'll leave a different impression than if it's in a big superstore. Not that one is right and one is wrong; it all depends on what fits with your proposition.

- Find good distributors who serve the right kinds of independent stores in the right geographical areas. Don't expect them to do all the work – get them enthused, to buy into what you're doing, and to partner with you on sales.

- Don't underestimate the importance and time it takes to service existing outlets. It won't be possible to spend your whole time bringing on new stockists. It's much harder than you think just to maintain a stockist:

every week, they're approached by new brands promising them the sales opportunity of a lifetime. Make sure you get in touch regularly to check they're happy, well-stocked-up with your product, and have all the point of sale they need.

- Track, learn and adjust. It takes time to get the processes right, but tracking sales data is a must. Understand which types of outlet are working well and which are not, and adjust your strategy accordingly.

About looking to the skies

Airlines are an interesting distribution channel and marketing opportunity that MOMA! makes full use of. The margins are tight, but the exposure to new customers compensates.

SECTION 3: BE UNIQUE AND SHOUT ABOUT IT

AVOID SAMENESS

Keep hunting for a manufacturer who isn't supplying all your competitors. It's hard work, but it will pay off in making you different.

If you have to no choice but to share a manufacturer with other brands, you MUST make sure your products taste different from the competitors' products that are also being created there.

You MUST taste different.

If customers can't tell the difference between your product and another product created in the same factory, you're not special, you're not unique and you'll have to compete for those customers on price. That's never a good situation to be in.

Always experiment. Always make new prototypes. Develop new flavours before anyone else does. Keep moving. Remember: own-label copycats are just behind you…

The Innocent drinks people still develop new flavours in their test kitchen at Fruit Towers.

Camilla at Higgidy Pies experiments in her own kitchen when she's working on new pies.

BANG ON ABOUT HOW YOU'RE DIFFERENT

If you can't tell me HOW you're different and distinctive, there's a pretty good chance you're not.

And the customer won't see you as different either.

Start here:

You have a great product, which tastes a whole lot better than the norm. How and why does it taste better? Tell people!

Innocent drinks told people how they were different – their smoothies aren't from concentrate, they're 100% pure fruit, and they contain no added sugar. And as a result they taste far more delicious than anything else out there, and they're super healthy.

Customers got it. They understood. They bought. They loved. And they told all their friends.

Stay different.

Don't devalue or change your product as soon as you get listings because of pressure on margins – it's not a good excuse. The trick is to stay high-quality and wait for savings to come as you sell in volume and can negotiate with ingredients suppliers.

STAY CLOSE TO THE MAKING

Most producers I know start by doing it by hand, and then get a kitchen.

If you're outsourcing from the start, it pays to learn how to make the product yourself **as well**.

It's the most useful crash course you'll ever do.

You'll understand flavour, texture, ingredients and shelf life.

Making will give you the language to understand and specify to an outside kitchen or factory exactly how you want your product made.

If you know the ingredients of your product, then you'll understand the impact on the flavour of changing those ingredients.

And you'll be able to argue your case better with your production people.

TELL YOUR STORY, EXPLAIN YOUR MISSION

"The greatest job of packaging is to make you want it before you even know what the thing is." Richard Reed, innocent drinks

Most food companies spend a lot of time thinking – quite rightly – about the quality of their product, the design of their pack, and the description of the food/drink. But they often miss out on a key aspect of their brand: their story and their mission. The author Simon Sinek calls this your "why".

If you can't explain *why* your product exists and what you hope it'll achieve for those who buy it, you won't connect with your potential customers, and you won't inspire them to "join" you on your mission, buy from you or use you over the (possibly cheaper) competition.

Does your packaging explain your values, and *why* you exist as a company? When people "get" you and are on board with you, they'll want to learn anything else about who you are and what you do.

Your values need to be clearly and immediately apparent. And you need to put a lot of effort into working on the words that express those values. (If you use meaningless, overused words like "natural" and "healthy", you won't get anywhere.)

Once you've explained your "why" (your story and your mission), your "how" (how you created your product to fulfil your mission) will have more impact.

The one-pot meal brand Easy Bean has a "Beanifesto", which states that it aims to "trade fairly" and "care for our environment" (among other missions to provide healthy, nutritious and delicious food). And *how* does it meet those aims? To trade fairly, they use local suppliers as much as possible, and they work with the Fairtrade Foundation to establish Fairtrade accreditation for pulses. And to help care for the environment, they've used polypropylene pots that can be recycled or reused; their sleeves are recyclable; and they use cutlery made from corn starch when holding tastings.

Now you feel *really* good about eating their food, and you probably don't even mind paying a bit extra for your lunch.

Or how about MOMA! They're "on a mission to revolutionise breakfast on the go and banish bad breakfast habits". And they know that "it's all too easy to kick-start your day with sugar, chemicals and grease", and that "grabbing the closest, quickest, most convenient thing is the order of the hour". So Tom Mercer, founder and farmer's son, created a wholesome, tasty and fuss-free breakfast that's filled with high-fibre, low GI jumbo oats, low-fat probiotic yoghurt and real fruit.

See how the "how" has so much more impact once the "why" has already been explained?

FINDING YOUR [TONE OF] VOICE

Just as your brand needs to have a mission and a story, it needs to have a tone of voice. But it needs its own voice – not something copied from an already-successful brand.

I'm talking mainly about innocent drinks: the chatty tone of voice that worked so brilliantly for them has been over-used and over-imitated. And for the most part, the imitation doesn't work because a) it's a poorer imitation, and b) it doesn't suit the brand of the product it's attached to.

So find your own voice. And don't let your designer or anyone else "do an innocent drinks" on you.

How's this for a great tone of voice (from a box of Flint & Co. tea):

> "Hello, my name is Ed Flint and I am a vicar. All my tea is loose-leaf, great quality and ethically sourced. I promise it will always be like that. And you can trust me; I have to report to God…"

It's funny, it has personality, it has a great product promise and it's different.

DO YOU HAVE AN INTERESTING BACKSTORY? USE IT!

Did you leave a high-paying City job to follow your passion for dips?

Are you a chef who thought it would be great to cook up a bacon jam?

Did you quit a promising career in neuroscience to become an eel-smoker?

Are you a vicar selling tea blends in Shepherds Bush?

Are you a former aromatherapist and Masterchef finalist turned chocolatier?

If you've got a great story that shows your passion for your product, use it. It's what people will connect with emotionally, before they've even tasted your product.

Amelia Rope's aromatherapy training comes straight through in her Pale Lemon and Sea Salt chocolate bar – it's the discerning trained "nose" and palate of Amelia that make all her flavours so distinctive and memorable.

YOUR PERSONALITY EXTENDS BEYOND THE PACK

People who like your brand personality will buy into your brand and – most importantly –buy your product. They'll go with you against the competition, because they associate themselves with you.

All successful food and drink brands have a personality. Think of innocent drinks (masters of pack chat), Rude Health (outspoken, even a bit ranty), Muddy Boots Foods ("countrypreneurial").

The personality of these brands is consistent throughout everything they do – their pack designs, their pack text, their websites, their adverts, their social media… everything. The pack is only one place for their personality to shine.

If you don't have a consistent personality, you won't seem authentic.

Let's say someone finds you via Facebook. Your presence there is dull and dreary, but your packs are vibrant and eccentric. That person won't get a good understanding of your brand, and when they're next in the supermarket, they won't recognise your product on the shelves.

But it's not just about Facebook, Twitter and Pinterest.

Be inventive. Think about what your brand "would do" to show off its personality.

Some examples:

- Muddy Boots Foods has videos on YouTube – the most irresistible of which is Miranda's husband, Roland, naked on a tractor, shouting: "Sex sells!" And the out-takes are even funnier.

- Innocent does knitted hats on bottles for AgeUK's The Big Knit.

- Firefly Tonics puts customer photos on their bottles.

- Higgidy offers a free (and rather elegant) apron if you send them a pie poem via their website.

Stuff like this sets you apart from the competition and makes people love you.

TELL US WHAT YOUR DAY IS LIKE

I ask shoppers in-store what they know about small food and drink companies. And most have no idea how small companies work so hard to get listed in a store, and how hard they work to stay there.

Does it matter? YES!

We're intrigued about how our food gets made, and the effort that goes into it. (Added bonus: when you show how hard you work, it makes your product seem like excellent value.)

So tell us. Send out a tweet when you've received a new delivery of hard-to-source very special West Ugandan Fairtrade vanilla pods. Post a Facebook picture of the new flapjack you've created the recipe for. Blog about your problems with the new super-powerful mixer.

Check out The One Mile Bakery's Facebook page for a brilliant example of communicating.

Cat Lyne of Cat & the Cream is reinventing cakes with her delicious, entirely vegan, wheat-free and soya-free cakes – completely handmade in her "artisan cakery" in Battersea.

She tweets and adds Facebook pictures of her ingredients going into the mixer, and the cakes on the cooling racks.

She has an enthusiastic and vociferous following on Twitter – people who love the flavour and the thought that goes into these high-quality, beautifully iced, individual indulgent treat cupcakes.

She's found a real niche and she takes her customers with her on her food producer adventure.

.

TIPS FROM AN EXPERT: USING SOCIAL MEDIA

Elisabeth Mahoney of The One Mile Bakery in Cardiff really knows how to work her social media.

She says:

Use social media before launch day

Use social media (Twitter, Facebook, Pinterest) to connect with people while you work towards launch. Think local and like-minded: find people, companies, food journalists, bloggers, campaigners, etc. close to your area of business or product.

Connect voraciously and generously: get chatting, share experiences, help out with requests for info, have a laugh, talk about lots of things other than your business. You can do all of this before mentioning your product/business – just identify yourself as someone interested in your field and find lots of people who share your interest.

Highlight your USP(s) and be yourself

You need to make clear what you're doing differently/better than others, but just as important is conveying your values, passions, personality, sense of humour and knowledge: it's these things that will bring people to your product or brand on social media.

Vary your social media output

Sure, some of your social media content will be purely promoting your product, but mix those with ones showing behind the scenes, the product in the making, etc. Use plenty of images and do so with a bit of engaging fun: I have a "bread of the week" tweet each Wednesday – my best-looking loaf – and it's called #breadpinup. The response to this is amazing every single week.

What do your favourite social media brands do well?

If you use social media as a consumer, think about the brands you enjoy engaging with. Why is that? What do they do well? Learn from that, and adapt good practice for your company. Which brands don't you engage with? Why? Avoid their mistakes.

Have something different to say

There'll be a lot of days, weeks and months to fill with your content, so be sure to spread out news/developments so that people get a sense of your story unfolding and become interested.

I have a rule of something small and new every week; something bigger and new every month. In my first six months of trading, I ran a range of competitions on the launch anniversary and did so entirely through social media. This brought a huge following.

Get your staff to use social media for you

Train them, show them, and encourage them: people love to meet the team once they've connected with your brand.

Make it part of your daily routine

Tweet at least once a day – at different times, about different things. And update your Facebook and Pinterest pages too. Also take time to chat to customers and people you've connected with – reply to their tweets about their day, etc. Five or ten minutes of this a day pays huge dividends.

A positive tweet or Facebook comment from one of your customers is marketing gold dust. Engage with them first ("So glad you enjoyed", etc.), then re-tweet.

Search on Twitter

Use Twitter to find local food bloggers and relevant (e.g. features, food) journalists in the local media. Build up a relationship with them – e.g. retweet a blogger's new post to your Twitter followers; tell them how good it was. Later, invite them to sample/review your product or come to a launch.

Try to get reviewed by local bloggers

Being reviewed by six local bloggers drove my business almost exclusively in the first three months, and I still get customers through those reviews. Local journalists read them too, and that's how I got my first in-print local coverage – which in turn led to a huge surge in business and coverage in a national paper (the Guardian's Cook supplement) within eight months of launching a tiny hyper-local company.

INNOVATE OR BE DE-LISTED

In the supermarket, nothing stands still. There are keen new entrants besieging the buyers ALL THE TIME and companies get de-listed ALL THE TIME.

To stay on the shelf, you'll need to do the following:

- Create sufficient and consistent customer awareness and demand for your products over time

- Have a decent rate of sale

- Refresh and renew products

- Come up with new products

Supermarkets have a tendency to steal ideas and do them under own label as soon as they judge that you've successfully trail-blazed an opportunity. You can't stop this, but you can anticipate that it will happen.

Innovate intelligently.

Entrepreneurs love innovation, and they tend to rush into making new products and new flavours.

But beware "time suck" on developing flavours that might intrigue you, but actually don't sell very well.

Keep an eye on the sales and see what's doing well… innovate based on the sales you've already got.

YOU HAVE FREEDOM, SO USE IT

The entrepreneur's secret is freedom.

Running your own show is completely and entirely addictive, and it's also a highly creative activity: you get to paint your own picture of who you are and share it with the world.

All the time. And in your own way.

No one defines what you need to do each day or helps you prioritise items on that to-do list.

Scary? Of course.

Magnificent? TOTALLY.

Being an entrepreneur means you can try stuff out. If it works, hurrah! If not, probably not that many people noticed anyway, and you'll have learnt something.

You can be like Jim Cregan of Jimmy's Iced Coffee, with his entertaining and bonkers "Make A White Russian In The Pouring Rain" YouTube video.

You can pitch to Nick Clegg, as Tom Mercer from MOMA! did with his "VOTE FOR OATS" campaign. And of course, you can film the entire thing.

You'll never be bored at your desk again. Yes you'll be overworked, challenged and tired. But you'll be exhilarated and never bored.

NURTURE THE BANKERS!

By which I don't mean HSBC or RBS, but the products in your range that sell really consistently.

After a while on the shelf, your sales figures will show you your "bankers": steady and reliable, performing consistently, non-dramatic.

Look after them and invest in them.

By all means, innovate and create new products or ranges or flavours. But don't get bored by or neglect your bankers.

Innocent drinks' top-selling smoothie has always been strawberry and banana, so they invested in it by making it better and adding in more strawberries. It was an excellent idea, because as competitors realised its popularity and rushed in to imitate it, innocent was already one step ahead (yet again) when it came to taste and quality.

SECTION 4: YOUR PACK DOES MORE THAN HOLD YOUR PRODUCT

SHAKE UP THE PACK

What's going on in your category when it comes to pack structure?

Are the other gourmet soups in pots? Are the "adult soft drinks" in glass bottles? Are all the other cereals in rectangular boxes?

Be different and stand out from them. Here are some tips on how:

Look at different categories and see how they're doing things

If you're selling soup, see how baby food is packed up. If you're selling smoothies, look at how milk is sold – or ketchup, or olive oil. It sounds crazy and most ideas won't work, but you might hit upon a gem.

Look to different markets

Go to a French supermarket and see what they do with their chocolate, their crisps, etc.

Think about different shapes

A different bottle shape can make you really stand out.

Bottle Green Drinks cordial is a good example of an elegant distinctive shape.

Toblerone owns "triangular" and riffs on it in limited edition packs to great effect.

Ella's Kitchen baby food pouches have incredible impact in-store – they are shiny, and colourful, and tactile – toddlers love holding them.

If people tell you, "That's not how X product is packaged – we can't do that," challenge them, keep pushing and ignore their objections.

Shoppers love to see and get a feel for good food

If your muesli looks delicious, make a window in your pack to show it off.

If your juice is a glorious (and natural-looking) colour, brilliant. People "eat with their eyes" and they use colour to judge the naturalness of a product.

Shoppers adore windows in packaging – they like to see the product. They also like to get a feel for it – they'll "squeeze" your flapjacks to see how firm they are, they'll shake your soup to figure out if it's thick or thin (and judge the value), and they'll feel the weight of EVERYTHING in their hand to tell how much they're getting.

THINK ABOUT YOUR PACK MATERIALS

If you choose different pack materials, structures and shapes from other products in your category, it'll help you stand out. But there are other benefits to choosing different pack materials…

Pouches instead of pots

How about pouches for soup instead of pots? Pots are energy-inefficient when you're hot-filling them on a production line and then blast chilling them – because they take longer to cool than pouches do. Also, they use a lot of plastic. Pouches have less material for recycling, they'll cost you less, and they chill quicker.

Recycled plastic

Innocent used recycled plastic in their little smoothie bottles, and it's definitely an eco-win with concerned shoppers. Recycled plastic is not as rigid on a filling line so that will limit the percentage of it you can use, but it's definitely the way to go.

Supermarkets hate glass

Avoid glass bottles for drinks. Supermarkets hate glass: it's dangerous and it's heavy. Consumers aren't that keen on it in their handbags either.

Tetra Paks are good for a longer shelf life as no light gets to the product (as it does in a bottle). Tetra Prismas look modern and are good to hold.

Tetras give you lots of space for branding, which a label on a bottle doesn't. (And they can hide your product if it doesn't look especially appetising! I once tested a pee-coloured drink that was packed in a bottle that looked like a hospital sample…)

TIPS FROM AN EXPERT: PACKAGING

David Myerson of Hurricane Design has developed brands through structural packaging for Unilever, Diageo, Danone and HP Foods.

Here are his tips:

Becoming an icon takes time

It's not just about developing a unique shape that achieves an instant success. You also need to stick by it for long enough to enable the shape to become an icon.

Be a product, not a pack

A successful pack can be a product in its own right. A pack for Derwent Pencils was so successful that it's now sold separately as a canvas pencil wrap – a new product line for the client.

Who uses me?

Watching your customer is essential. Ever watched a team of 15-year-old boys hurling a drink around a rugby pitch? We did. So the Vimto bottle has curved contours for easy grip, ridges at the neck and a sports cap to make throwing and drinking "on-the-go" easier for the brand's teen target audience.

KISS

Keep It Simple Stupid.

Sometimes it just needs simplicity, great texture, colour and materials.

(So look at the sheen on your cardboard – it can make all the difference!)

Find out more about Hurricane Design:
www.hurricanedesign.com.

YOUR PACK SIZE IS PROBABLY TOO BIG

The most common problem I see in research is companies using packs that are actually **too big** for their target market's needs:

- Too much to eat before the use-by date

- Too generously sized – leaving not enough profit for the company

I've interviewed *many* shoppers in the aisle, and what they say they hate most is waste – wasting their money on too large a pot of something that they can't eat in time before it passes its use-by date.

When the unused product gets tossed in the bin, it's a double waste – of their money, and the energy and time YOU invested in creating it, distributing it, and marketing it.

I also know from my research that the most common comment from shoppers is: "It's too expensive; can it be cheaper?"

"Can it be cheaper?" is the most common feedback that innocent drinks get. People in the UK expect their food to be cheap, and supermarkets with incessant price promotions encourage them in this belief.

Kill two birds with one stone!

Size your product down and charge less. Give them just about enough of your product, and stop there. Doing so also allows you to get a good balance between your costly ingredients and margin, and people's idea of good value.

If you decide to work your way up from the independents, it's easy to be too generous because you'll be making more of a margin. But don't get sucked in, because when you want to move to supermarkets, they'll demand more of your margin.

SMALL PACKS ARE IN FASHION

In these economically straightened times, small is beautiful in the eyes of the shopper.

And smaller has always signalled fewer calories to women – so a small dessert or chocolate bar becomes an acceptably sized treat (see Green & Black's mini bars, or G'NOSH dippables).

And there are a number of brands who've recently downsized their product but kept it at the same price – particularly because of increases in raw material costs, but also because customers are exceptionally price-resistant in some supermarkets.

M&S have a teeny tiny packet of prosciutto at £1 (and Waitrose have two slices for £1.05.) It's a perfect little treat and means NO waste on the part of the consumer.

There's been huge success over at Gü Puds with their relatively small but very delicious "shot-style" high-quality puds.

EXPENSIVE AGENCIES NOT REQUIRED

The moment you move from the market stall to the aisles, you can't do the talking and your packaging has to do it for you.

And when start-ups reach the supermarket stage, they get *very* excited about branding and design. Because it's fun, it's sexy, and it's much more interesting than distribution, manufacturing and "hazard analysis and critical control points".

And then all the excited start-ups go to the five expensive design agencies that everyone else goes to see.

Stop and think for a moment!

- You don't have to use these really expensive agencies.

How about using a newer, keener, different design agency?

There are loads of benefits in doing so:

You'll differentiate yourself

You DON'T want to be like all the other packs out there. If you've got an amazing, delicious, different product, why package it up so that it looks like all the others? Why would anyone choose **you** over an established brand if they think you're no different, they don't know who you are yet, and you're possibly more expensive?

Each design agency has a particular style and way of doing things – which means that lots of the companies they work for have very similar styles and tones of voice.

Take Boden and Dorset Cereals for example. I bet you can tell how similar their branding and tone of voice is. That's because both came from the same design agency. It's a very particular style, and it may not be right for you. But more importantly, *you don't want to be like all the others anyway – you want to stand out.*

You'll save money

Big, established design agencies cost a fortune. So if you go with someone newer, you'll probably be spending far less on design – which means you can spend that money on equally important priorities like sampling, marketing and getting listed.

You'll get more of their time and effort

With big, established design agencies, it'll be hard to get the time and attention of the people at the top. That won't be so much of a problem with newer, keener agencies.

What's more, they'll be more invested in your success: if your product isn't doing well, they might not get that much-needed testimonial or promise of further work with you. Big agencies care less, because they're inundated with work and don't even need testimonials anymore.

All the above points apply to your website too: don't go for a big, established web-design agency. There are plenty of *fantastic*, original, keen and capable small web-design agencies or freelancers out there. Give them a try.

PACK COLOURS AND DESIGNS MUST ACCURATELY REFLECT YOUR BRAND

A good designer can communicate anything you want with the right combination of colours, graphics and font size, and a symbol, window, drawing or photo of your product.

So what do you want to say?

Camilla Barnard of Rude Health has deliberately made their beetroot and pumpkin snack bars really stand out:

> "Bright colours suit our [brand] personality. We are outspoken, modern and energetic. Our colour palette should reflect this – contrasted with the craft background that shows our ethical side."

When you open up the simple brown wrapper of an Amelia Rope Chocolate, you'll find a beautiful, metallic, gleaming, crisp-coloured foil underneath. It increases the sense of specialness and anticipation when you open one, and it perfectly suits the understated luxury of the brand.

Christina Baskerville of Easy Bean has created a simple bean logo against a strong coloured background. It's immediately obvious that it's all about the beans here – which helps draw in vegetarian customers (for whom she's also added a "V" to the pack), and slimmers (for whom she's added the number of calories to the pack). Ingredients run along beneath: chickpea, aubergine and butternut squash. And at the top it says "One-Pot Meal". It's simple, effective, easy to understand and brilliant.

But remember: even if you're an understated brand, your product needs to stand out on the shelves. Don't use obscure pale fonts on white backgrounds unless you want to sink without a trace.

The shopper is busy, so you need to help out by being very clear about what your product is and why it will appeal, and making your font easy to read from a distance.

THE PLACEMENT OF YOUR PACK MESSAGING

You don't control how your product is displayed.

Stacked packs

If you have a sleeved pack, any supermarket will probably display it stacked. That means that unless shoppers pick it up and really examine it, **they won't see the messages on the top of your pack.**

Shoppers don't scrutinise products unless they're massive foodies or they have intolerances.

So when you brief your designer on your cardboard sleeves, make sure they think about what it communicates **from the sides.**

Refrigerator aisles

I watch people at the yoghurt fixture, the drinks aisle, the soups aisle... these places are **cold.**

People don't hang about in refrigerator aisles for long, and they don't touch products much. They look with their eyes, select quickly, and move on.

So think about whether your chilled product works from some distance away.

Is it giving enough information to someone in a hurry? What's it telling them?

Test your assumptions: get your designer to mock you up some packs. (If the designer squeaks because they want the job signed off, just say "Not quite yet.")

Then sneak some packs onto the supermarket shelf – as James Averdieck did with his Gü Puds – and see if people notice them. (I really suggest chatting up the supermarket manager if you're going to do this – as I do when testing out products. If you have a friendly supermarket manager where you're stocked, explain why and how you want to do this – it all comes down to the relationships you've cultivated!)

Oh look... bingo! You know your pack works if they pick it up – this is what you're aiming for. If they look at it and move on, find out why they put it back – what was it NOT telling them?

HELP SHOPPERS BUY INTO YOU: PROVIDE "TASTER" SIZES

Different pack sizes work for different eating occasions and allow people to buy affordably into your product. Which is essential, when you're new. You can move them up your chain of products – the key is to get them started on buying you.

Peter's Yard crispbread packs choices work like this.

The crispbreads come in a "canapé" size, an "everyday" size, a large "wheel" sharing size, and in a variety of tins for storage and gifting. They can be eaten as canapés at parties and casually for lunch. The smallest size is the perfect entry and price point for people new to the brand.

G'NOSH dips come in a snacking size – "GNOSH dippables", with breadsticks made by Crosta and Mollica – which is a handy smaller version alongside their bigger family pack.

GET INTO GIFTING

Speciality chocolatiers thrive on the big sales hikes of Valentine's Day, Easter and Christmas. But gift-giving isn't restricted to those big, national days. Birthdays, dinner parties, anniversaries… gifts are needed year-round.

Think about whether your product can be considered a gift as well as a regular supermarket item: it'll open you up to a whole new market.

The pack matters though: it needs to look like a gift if it's going to be thought of as one. I know from my work with Cadbury and researching Easter packaging for them that a gift box needs to have a certain elegance.

My gluten- and dairy-intolerant friend's eyes lit up when I took her Seggiano's Almond Chocolate Torte for dinner. And my dad loves to receive the beautifully packaged Sipsmith Sloe Gin.

Some surprising product categories have gift potential. Take the elegantly attired mid-50s woman I met in John Lewis Food Hall; she was intrigued by the Peter's Yard crispbreads in a flat box in a giant wheel shape.

She was going to take it to a cheese-loving friend for lunch, because "She'll be really intrigued, she loves crisp cheese biscuits, the packaging is Swedish and lovely, and it's only £5 and a great little gift."

She could have taken flowers, but her friend was a foodie and the crispbreads were stylish enough to work as a signifier of her personal taste.

You can see from the above that you need to get all the elements right for this to work: packaging, price, authenticity, novelty and quality.

SECTION 5: THE DREADED SHELF LIFE

THE PROBLEM WITH PERISHABLE

Popcorn and crispbread makers have it (relatively) easy – and so will you if you've chosen to make a product without any perishable ingredients and with a longer shelf life. Your major issues will probably be around distributing fragile products – and you'll soon be sick of the sound of popping bubble wrap.

But if you've chosen to make a chilled product from fresh ingredients, things are quite a bit tougher... because fresh ingredients degrade over time, and apart from gas-flushing and packing in a protected atmosphere, there isn't – yet – that much you can do to slow down perishing rates. New packaging technologies are coming on-stream from Kerry Group's Innovation Centre, which may become more affordable in the future.

Recommendation: if you're producing a drink, go down the pasteurisation route. The raw milk brigade won't approve, but microbial growth is *not* something you want to be worrying about.

And be *very* careful about hygiene standards in baby food.

THE PROBLEM WITH SEASONAL

Think about seasonality.

Some products – like soup, burgers and iced coffee – are much more popular at certain times of year. To a lesser extent, the same goes for chocolate – Christmas, Easter, Valentine's Day…

You may think you can change shoppers' habits in favour of your chilled drink in winter, or chilled soup in summer, but it'll be a real uphill struggle.

Are you really happy (and financially prepared) to have one or two "high seasons" followed by quiet periods?

Ideally you want to be selling year-round, which means you need to be thinking about year-round products (or building a business model where you have summer and winter products to even out your cash flow and support your product kitchen).

GIVE YOUR SHELF-LIFE SOME BREATHING SPACE

If you product is perishable, make sure it has a shelf life of over ten days. Do NOT go for an eight-day shelf life. Here's why:

- You'll lose a day in distribution, so your product will only be on the shelf for a week. And remember: it's a week on the shelf *only if* the supermarket gets you out of the storeroom/depot and onto the shelf smartish. They may well not...

- The food purists won't love you for a ten-day shelf life, but they're more likely to be visiting their organic deli than the supermarket.

- Importantly the retailer *will* love you (as well as the organic deli actually, because you'll provide them with something that can stay on the shelves a while).

- You're less likely to suffer the pain of seeing your products marked down – a frequently repeated rite of passage for makers of eight-day products.

- For an eight-day shelf life, your rate of sale (how fast you sell) will need to be high – and that's unlikely to happen when you start out (unless you have LOTS of help with sampling your product in-store, or the owner of the store offers to help recommend you).

So right from the start, do yourself and your stockists a favour: make that shelf life a bit longer if you can. It's not selling out. It's selling IN to the shops where you need to be stocked.

You don't need preservatives for products to last ten days. It's all about the packaging. Go and look at other products in the supermarket, inspect their packaging carefully, and steal their secrets.

SECTION 6: RESEARCH RESEARCH RESEARCH

TWO EASY RESEARCH METHODS

Family and friends will tell you it's great. But they love you and they're biased, so you can't just test your product on them.

And you can't just try it out on people who're "just like you" either – it's not representative enough of the population.

Innocent drinks made a veg pot with edamame beans in it, which went down fine in their London office (where everyone knew what edamame beans were), but did FAR less well in tests in Bradford (where people didn't know what they were eating).

You need to get feedback from people you don't know, and people who aren't just like you: *these people* need to find your idea irresistible and want to buy your product.

You don't need to spend too much time and research on research. Here are two easy methods for you:

Research method 1: Take your products to new audiences and test them

If you have a friend who works in a large company, ask if you can bring in some food/drinks to test on their colleagues. You'll need to spend some money on samples, but that's infinitely better than launching an imperfect product that doesn't sell.

Lucy Thomas, former innovation manager at innocent drinks, explains further:

> "Find out where your potential customers are and tap your contacts to help you get in front of them.
>
> Do a taste test in a friend's office; people are usually receptive if they're offered something back.
>
> Test on as many people as possible. Ideally 100, but 40–60 will give you a good steer. Encourage them to be as honest as possible.
>
> Test against your competition. Decant all the samples into unbranded packaging, to avoid bias.
>
> 11 am is the optimum time for tasting, apart from chocolate – which is best tasted first thing.
>
> Be clear on what targets you want to achieve before you do the taste test.
>
> On a 'just about right' scale (e.g. for sweetness) innocent would expect at least 60% of people saying 'just about right'.
>
> Look at the distribution of scores.
>
> Even if you hit your average, you might have a polarising product.

That's not necessarily a bad thing, but like Marmite, if people either love you or hate you, it's going to limit your market."

You could always let people take your product home...

It's often very enlightening to see what people do with your product over time. So if you can, ask them to take your product home for a week.

Ask them to make a note of every time they eat it, how they eat it, why they eat it.

Then send them an email and ask what they did with your sample.

If they ask for more, you're doing things right.

If they have lots of criticisms, it'll hurt but it'll be useful: far better to fix your product now than after it's failed on the shelves and been de-listed.

So make sure you PAY ATTENTION to their criticisms. You may not agree – and you may not do anything about criticism if enough people offer the opposite viewpoint – but you need to listen to them all and evaluate objectively.

You want a lot of "Ooooh!" and "Yum!" before you can be satisfied that your product is perfect.

If you want to benchmark your product against others, try Wirral Sensory Testing – a sensory and consumer research company that tests your product against others in your market with consumers. They tell you how you compare, and what you need to change in your product, e.g. sweetness levels.

It's not cheap though, so it's best if you have investors' money to spend on it.

www.wssintl.com

Research method 2: Get a market stall

Get that market stall as soon as you can. It's low cost, and it gets you out there.

You'll start getting useful feedback from day one – which products sell, what people understand (and comment on!) about your branding and ingredients story, the price, and the size.

This is market research for free.

You'll hone your sales skills – learning the best ways to explain the why and what of your products.

You'll see which of your products appeal.

You'll get a sense of why and how they work from the reactions and comments you get as people taste your samples.

Your sales over the weeks or months are the ONLY indicator of what works.

LISTEN HARD, and think about what your customers are telling you.

And you might hear a free strapline too!

You'll find that there's a short-cut sentence or phrase or description that really resonates with them and explains you fast.

Claire Martin is the founder of Breckland Orchard, which she started to call "posh pop" to explain her positioning among competitors. It fits so well that she's bought the rights to use it.

This phrase describes her soft drinks perfectly. Work on finding yours...

Be prepared to "pivot" if your research tells you to.

It's very easy to be wedded to the first idea you started with, but customer interest and sales might tell you to refine that idea. You need to be prepared to listen.

Muddy Boots Foods started out making ready meals, but then they found that the real demand was for their burgers.

And here's a tip to keep your feet warm (well, warm*er*) at markets:

Always stand on cardboard. (Thanks to Claire Martin of Breckland Orchard.)

GIVE IT AWAY TO GET LOADS MORE BACK

I'm talking about sampling here. And all the successful food brands do it.

Innocent drinks carpet-bombed the BBC with smoothies in their early days, so that every journalist there knew about them.

Why sample?

It's "edible advertising"

Sampling earns you real attention from real people. Attention: a commodity in ever-shorter supply these days.

It's the best way for unknown brands to get known. **Think of it as edible advertising.**

As Tom Mercer of MOMA! explains:

> "Sampling is our key to getting consumers into the MOMA! world.
>
> The main thing is getting people to physically try our breakfast – our product is pretty unique so it's really important for people to taste it in order to understand and appreciate it – and once we've done that most people love it. Quite simply, it's the best way for us to introduce the product to people and encourage purchase afterwards."

Your customers will get into the habit of tasting your product

If you're recently listed in a supermarket and sales aren't going great, regular sampling will get your sales up and keep them up, because shoppers will form a habit of trying out your product each time they go to the supermarket.

You'll create an obligation to buy

Sampling can get you in front of customers who then get to know you. And that creates an obligation (in some people's heads – not all!) to buy…

How to sample?

In a nutshell: sample your arse off, all day, every day, and keep sampling. Get yourself out there and make a noise.

In more detail…

You need confidence, in-depth product knowledge, and HEAPS of enthusiasm.

Confidence encourages people to stop what they're doing and take a sample. But then you need good product knowledge to help the shopper understand what it is they're trying – this can often inform their reaction to it.

You also need heaps of enthusiasm for your product – it'll transmit to the shopper and helps to get a great reaction from them.

As Tom Mercer from MOMA! says:

> "Products genuinely taste better when people are lifted with great brand energy!
>
> If done well, sampling not only converts people to your product and results in more sales, but it also creates a great brand image. We want people to come away from trying a MOMA! sample thinking how great our product is – and part of that is an encounter with us."

When to sample?

Only through regular sampling slots will you discover your best day/time.

Miranda Ballard of Muddy Boots Foods has done so much in-store sampling that she knows the exact time that supermarket shoppers are most ready for a sample of freshly cooked burger: 11.38am.

If you're sampling in the independent stores rather than supermarkets, it's probably worth asking them in advance when they're busiest: slow days can be tedious and dispiriting.

Some stores will only allow sampling at certain days and times – so be sure to check with them.

For example, Whole Foods actively encourages tastings on their busiest days – Saturdays (between 11am and 2pm) – so that shoppers can get to know the specialist food producers they stock, like Amelia Rope Chocolates and Cat & the Cream's vegan cupcakes. Both Amelia and Cat will testify that sampling really builds business.

BE BOLD, BE A SALESMAN

You're a salesman now, so you need to think and act like a salesman (no matter how nervous and out of your comfort zone you're actually feeling).

So... **be bold:**

- Barrel up to strangers and tell them about your great new product – over and over. It's very handy for sampling!

- Think creatively about how you can get in front of the buyers you need to meet.

- Take rejection in your stride – from stores and customers.

- Be comfortable with the unknown.

- Remember that if you've chosen to make something of quality, not everyone will care or be interested. Don't take this personally if you can't convert them.

- Find the tribe that **does** care and **is** interested in you.

- Thank them and ask them to spread the word!

- Block out negative feelings.

Extra tip on being bold: before you go to pitch to a supermarket, nip into the toilet and make yourself as big as you can. Stretch your arms up to the ceiling. Honestly, it's a proven method for increasing testosterone – especially for women. Deep breath in and out. Go sell with a smile!

AND FINALLY...

STAY DETERMINED

Every day you'll have highs and lows, set-backs and triumphs.

Stay determined.

Just keep on keeping going.

I've watched so many businesses graft away, and eventually they turn a corner, and it WORKS. It really does. And they can feel it. They've arrived!

To all of you, I salute you and I am with you!

In the words of the splendid Jim Cregan of Jimmy's Iced Coffee (whose Facebook page shows him reclining like a very humorous, determined, wet-suited seal)…

KEEP YOUR CHIN UP!

ACKNOWLEDGEMENTS

I want you all to know that I now have an extensive "chilled aisle" wardrobe.

Big thanks to you all:

Emma Jackson, MD of Brekfix, for taking a chance on me at innocent drinks

John Vincent of Leon Restaurants for suggesting I write a blog

Richard Reed, co-founder of innocent drinks, for projects and inspiration

Lucy Thomas for generous sharing

Charlotte Knight of G'NOSH

Tom Mercer of MOMA!

Camilla and Nick Barnard of Rude Health

Jamie Mitchell of Daylesford Organic

Jim Cregan of Jimmy's Iced Coffee

James Foottit of Higgidy

Dan Shrimpton of Peppersmith

Miranda and Roland Ballard of Muddy Boots Foods

Wendy Wilson-Bett of Peter's Yard

Harry Briggs and Marcus Waley-Cohen of Firefly Tonics

James Averdieck of Gü Puds

Chris Jones of Gourmet Raw

Lee Robertshaw of Together Drinks

Dana Elemara of Arganic

Bethany Eaton of Coyo

Cat Lyne of Cat & the Cream

Hiromi Stone of Kinomi London

Katie Christoffers of Matcha Chocolat

James Cronin of Paul A Young

Amelia Rope of Amelia Rope Chocolates

Silvana de Soissons of The Foodie Bugle

Christina Baskerville of Easy Bean

Lucy Wright and Anna Mackenzie

Alastair Instone of School of Food

Nishul Saperia and Tahzeen Basunia of Jealous Foods

WEBSITES TO VISIT

I've mentioned or quoted many founders from fantastic food companies. If you want to find out more about them, here are their websites:

Amelia Rope Chocolates: **www.ameliarope.com**

Bear Nibbles: **www.bearnibbles.co.uk**

Bottle Green Drinks: **www.bottlegreendrinks.com**

Breckland Orchard: **www.brecklandorchard.co.uk**

Cat & the Cream: **www.catandthecream.com**

COYO: **www.coyo.co.uk**

Daylesford Organic: **www.daylesford.org**

Easy Bean: **www.easybean.co.uk**

Flint Tea: **www.flint-tea.com**

G'NOSH: **www.gnosh.co.uk**

Graze: **www.graze.com**

Gü Puds: **www.gupuds.com**

Higgidy Pies: **www.higgidy.co.uk**

Innocent drinks: **www.innocentdrinks.co.uk**

Jamsmith: **www.jamsmith.co.uk**

Jealous Foods: **www.thejealouslife.com**

Jimmy's Iced Coffee: **www.jimmysicedcoffee.com**

MOMA!: **www.momafoods.co.uk**

Muddy Boots Foods: **www.muddybootsfoods.co.uk**

Peppersmith: **www.peppersmith.co.uk**

Peter's Yard: **www.petersyard.com**

Rude Health: **www.rudehealth.com**

School of Food: **www.cookeryschool.com**

Scratch Meals: **www.mealsfromscratch.co.uk**

Seggiano: **www.seggiano.com**

Sipsmith: **www.sipsmith.com**

The One Mile Bakery: **www.onemilebakery.com**

We Are Tea: **www.wearetea.com**

Yorkshire Provender: **www.yorkshireprovender.co.uk**

CPSIA information can be obtained at www.ICGtesting.com
Printed in the USA
LVOW08s2310230114

370707LV00001B/52/P